By George B. Stevenson

PRINTED IN   GREAT BRITAIN

DEAN & SON Ltd.

52,54 Southwark St.   LONDON SE1 1UA

TRADE MARK

CONTENTS

MULE DEER
fawn

603 00527 6

This edition Dean & Son Ltd, 1978

ROCKY MOUNTAIN GOAT and kid

Mammals are warm-blooded furry animals. No other animals have hair. Except for a few rare kinds that lay eggs, all mammals give birth to their young. The new born are fed milk from the female's mammary glands. Mammals range in size from the whales that weigh more than 100 tons—the largest animals that have ever lived— to tiny shrews that weigh less than an ounce when full grown.

Most mammals are adapted for the kind of life they lead. Thick fur or layers of fat (blubber) help to keep polar animals warm. Those that live in hot deserts need only small amounts of water to survive. Climbers use their limbs for grasping, and some have a tail to use also. Most swimmers have webbed feet, but in whales, the front legs have become flippers and the hind legs have disappeared. Bats, the only mammals capable of true flight, have wings of

POLAR BEAR
to 1,000 lbs.

ZEBRA
5 ft. at shoulder

thin skin stretched from their long fingers to their short hind legs. Diggers, like moles, have powerful front feet and claws. Zebras and other hoofed mammals escape from enemies because they are fast runners. Flesh eaters, like wild cats, have sharp, strong teeth. Plant eaters, such as buffaloes, have broad, flat teeth for grinding. A rodent's teeth are chisel-like for gnawing.

5

Bats

Mouse

Squirrel

Rabb

RODENTS

Scaly-ant-eater

Ape
Man

PRIMATES

Armadillo

Flying Lemur

Shrew

Ground Sloth
(extinct)

Mole

Duckbill

Opossum   Kangaroo   MARSUPIALS

All mammals are very much alike in genera
body plan. Yet each of the more than 4,000 kind
is different. The family tree above shows the

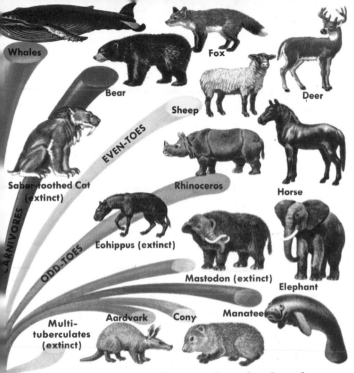

Whales

Bear

Fox

Deer

Sheep

EVEN-TOES

Saber-toothed Cat (extinct)

Rhinoceros

Horse

CARNIVORES

ODD-TOES

Eohippus (extinct)

Mastodon (extinct)

Elephant

Manatee

Multi-tuberculates (extinct)

Aardvark

Cony

main groups of mammals. It is thought that the ancestor of mammals may have been a flesh-eating reptile that lived some 200 million years ago.

POUCHED MAMMALS (marsupials) give birth to very small young. They stay in their mother's pouch until they are nearly grown.

The only pouched mammals in the Americas are the opossums. The Common Opossum, which may "play dead" when frightened, lives in woodlands from Canada to Colombia. It feeds on small mammals, birds, and fruit. The Water Opossum of Central and South America has webbed feet.

COMMON OPOSSUM
body 20 in., tail 10 in.

RED KANGAROO

Many kinds of pouched mammals live in Australia. Best known are the kangaroos. Most kinds are small. Rat Kangaroos are not as big as rabbits. Rock Wallabies are only three feet tall and Tree Kangaroos are even smaller. Largest kangaroos are the **Great Grey** and the **Red**, which grow six feet tall and weigh 200 pounds. A large male can jump 25 feet and can hop along at 25 miles an hour. Kangaroos eat grass or leaves.

KOALA
2 ft.

LESSER FLYING PHALANGER
19 in., including tail

The Koala is an Australian pouched mammal that looks like a small, cuddly teddy bear. This slow, peaceful climber eats only the leaves of eucalyptus trees. A full-grown Koala is about two feet long and may weigh 30 pounds.

Many of the pouched mammals found in Australia and other nearby islands look rather like mammals found in other parts of the world. The phalangers resemble flying squirrels; the Tasmanian Wolf looks like a wolf or a dog. There are also pouched moles, rats, and mice.

HEDGEHOGS, MOLES, AND SHREWS are a group of small mammals with tiny eyes and long snouts. Most are insect eaters. Hedgehogs of Europe and Asia have spines mixed with their coarse hair. Moles, found nearly everywhere in the world, live underground. Shrews are tiny animals found on all continents except Australia. The European Etruscan Shrew — 2½ inches long, including its tail — is the smallest of all mammals.

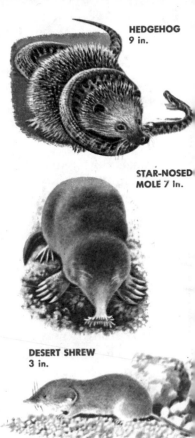

HEDGEHOG
9 in.

STAR-NOSED MOLE 7 in.

DESERT SHREW
3 in.

BIG-EARED BAT
4 in.

BATS are the only mammals with wings. The hairless skin of the wings stretches from the tips of the long fingers to the sides of the body and from the hind legs to the tail. Bats sleep during the day, hanging head down in caves or hollow trees. They become active soon after dark. In winter, those that live in cold climates either hibernate or fly south.

Most bats eat insects, which they catch in flight, but a Central American bat feeds on fish. Vampire bats, also of the American tropics, scrape skin off sleeping victims and then lap up the blood. One group of bats in the tropics eats fruit or gets nectar from flowers. All bats make high, shrill squeaking noises. These sounds echo back (like radar) from objects ahead. The bats are then able to avoid the obstruction. The sounds are too high-pitched for us to hear.

bats resting

LITTLE BROWN BAT
3½ in.

VAMPIRE BAT
3 in.

FISH-EATING BAT 4 in.—

Blue Whale compared in size to other familiar animals.

WHALES spend all their life in water but must surface to breathe. They give birth to their young at sea and nurse them while swimming. Whales are descendants of land animals. They do not have hind limbs. Their front legs have become flippers used in swimming. A whale "spouts" when it exhales warm air through its blowhole, the nostril on top of its head.

The Blue Whale, largest of the whales, grows to a length of 100 feet and a weight of 115 tons. It belongs to the group of whales that feed by straining tiny marine creatures through rows of whalebone, or baleen, on the upper jaw.

Toothed whales have teeth instead of baleen. The largest of this group is the Sperm Whale, 60 feet long. Dolphins, porpoises, the Narwhal, and the Killer Whale are toothed whales. The rare river dolphins of Asia and South America are the only whales that live in fresh water.

**KILLER WHALE
15-20 ft.**

RODENTS form the largest group of mammals. Their front teeth keep growing as long as the animal lives. Gnawing wears them down and keeps them sharp.

The Capybara (100 pounds or more) is the largest of all rodents; it lives in the marshes of South America. Another South American rodent is the Nutria, which escaped captivity and is now common in the U.S. The Guinea Pig and the Chinchilla are also rodents native to South America.

CAPYBARA
4 ft.

NUTRIA
body 22 in., tail 22 in.

BEAVER
body 27 in., tail 16 in.

PORCUPINE
body 28 in., tail 6 in.

Porcupines are rodents that are found in most parts of the world. They have quills mixed in with their fur. If an enemy gets too close, the porcupine makes a swipe, sticking the sharp quills into its attacker.

Beavers cut down trees to build dams and lodges. The entrance to the lodge is underwater, but the nest is above the water level.

KAIBAB SQUIRREL
19-21 in.

Squirrels of many kinds live on all continents except Australia. The largest, three feet long, are from India and Malaya.

KANGAROO MOUSE
5-7 in.

HOUSE MOUSE
5-7 in.

MICE, RATS, VOLES, AND LEMMINGS are the most abundant of all rodents. Most are small. The House Mouse, Black Rat, and Brown Rat have lived with man for centuries. The Meadow or Field Mouse is one of the small voles; the Muskrat of North America is the largest. Lemmings live near the Arctic.

BROWN LEMMING
5½ in.

COLLARED LEMMING
5½ in.

RABBITS, HARES, AND PIKAS are like rodents but have two pairs of front teeth rather than one. Rabbits are naked and blind at birth; hares are furry, have their eyes open, and can soon run around. Jack rabbits, which really are hares, can jump 20 feet.

Pikas or Conies are rabbit-like and appear to have no tail at all. They live high in the mountains of western North America and Asia, commonly above the timber line.

PIKA
7-8 in.

BLACK-TAILED JACK RABBIT
18-22 in.

COTTON-TAIL
12-15 in.

ELEPHANTS are the largest land animals. Their long, flexible trunk ends in a pair of nostrils. Their huge ivory tusks, which may weigh over 250 pounds, continue to grow throughout the elephant's life. A bull African Elephant may be almost 12 feet tall at the shoulder and weigh more than six tons. The African Elephant has larger ears and a longer, flatter head than the smaller Indian Elephant. The Indian Elephant is trained as a work animal. Both kinds live in herds that forage mostly at night. They eat leaves, fruit and small branches.

The Rock Hyrax, or Coney, of Africa and the Near East, is a small relative of elephants.

ROCK HYRAX
12 in. long

AFRICAN ELEPHANT
12 ft. tall

MUSTANG
wild horse of the
American West

HORSES, TAPIRS, AND RHINOCEROSES

are hoofed mammals with an odd number of toes on each foot. A horse actually runs on one toe. All its other toes have either disappeared entirely or are reduced to bony splints. Horses, native to Europe, Africa, and Asia, were introduced to North America by early explorers. The wild horses of Western America are descendants of horses that were turned loose or escaped. Zebras (p. 5), close relatives of horses, roam the plains of Africa.

Tapirs, of Malaysia and tropical America, have four toes on the front feet, three on the rear. Their short, flexible trunk is used to pull leaves from small trees and shrubs. They live in low country, usually near water, and are good swimmers.

Africa is the home of the Black Rhinoceros, the most numerous of these heavy mammals. The White Rhinoceros of Africa is the largest of all the rhinoceroses and may weigh four tons. The Indian Rhinoceros has only one horn, while the Black and the White have two horns. Rhinoceroses have thick hides and poor eyesight.

WHITE RHINOCEROS 14 ft. long, 5 ft. tall

PIGS come from the Old World but have been taken almost everywhere by man. They are even-toed, hoofed mammals (as are camels, deer, sheep, and goats). The Wild Boar is the only native pig in Europe. Africa has three kinds: the Red River Hog, the Giant Forest Hog and the Warthog.

The Hippopotamus, found only in Africa, may weigh four tons. Hippos feed on grass.

RED RIVER HOG
4½ ft.

HIPPOPOTAMUS
14 ft. long, 5 ft. tall

ALPACA
Body 4 ft.

CAMELS AND LLAMAS The one-humped Dromedary once roamed the deserts of Arabia and North Africa. A few wild Bactrian Camels are still found in Central Asia, but nearly all the camels of today are domesticated.

Llamas, which belong to the same family as camels, are used as pack animals in the mountains of South America. The Alpaca, as well as two other species of llamas, the Guanaco and the Vicuna, also live in South America.

25

DEER are hoofed mammals. None live in Africa or Australia. The Caribou of North America and the Reindeer of Europe live in the north and arctic regions. The Moose is the largest American deer and the Pudu of South America, only 15 inches tall at the shoulder, is the smallest. The Key Deer, found only in southern Florida, is also very small.

WILD CATTLE are large, horned, cud-chewing grazers. There are Wild Cattle on all continents except South America and Australia. The Yak, of Tibet, and the Water Buffalo, of Southeast Asia, have been tamed. The African Cape Buffalo is too fierce to be domesticated, and hunters are wary of it.

Bison, often called Buffalo, were once common. The European Bison, also called the Wisent, is slightly smaller and less shaggy than the more famous North American Bison. Both kinds now live only in game refuges and zoos.

KEY DEER
2 ft. at shoulder

MOOSE
6 ft. at shoulder

CAPE BUFFALO
5 ft. at shoulder

BIG-HORN SHEEP
3½ ft. at shoulder

**AOUDAD OR
BARBARY SHEEP**
3 ft. at shoulder

SHEEP AND GOATS
are native to Europe, Asia, North America, and North Africa. In North America the Big-horn is the wild sheep of the mountains. The slightly smaller Barbary Sheep is the wild sheep of Africa. The Mouflon, only wild sheep of Europe, is smaller still. Only the male sheep (ram) has horns, but both male and female goats have horns. Horns of sheep grow out to the side and then curve forward.

IBEX
2½ ft. at shoulder

Horns of goats grow up and then turn back.

Markhors, wild goats of Asia, are heavily built, with large, twisted horns. Tahrs, also from Asia, are smaller and beardless and have short horns. Several kinds of Ibex live in the mountains in Europe, North Africa, and Asia. Wild goats live in rugged, mountainous country. These sure-footed animals climb steep, rocky slopes where they feed on herbs, grasses and other small plants.

MUSK OX
6 ft. at shoulder

MUSK OXEN, ROCKY MOUNTAIN GOAT, AND CHAMOIS show features of both goats and antelopes. The Musk Ox lives in the Arctic where it feeds on moss and small plants. The Rocky Mountain Goat (p.3) is found in the Rockies from Montana to Alaska. The Chamois, once abundant in the mountains of Europe, has been hunted almost to extinction.

ANTELOPES are swift, graceful runners. Both males and females usually have horns. The Blackbuck and several kinds of gazelles live in Asia. Africa's herds of antelopes roam the plains and scrub forest. They may often be seen at waterholes. Some have long, straight horns; others have curved or twisted horns.

SABLE ANTELOPE
4½ ft. at shoulder

BLACKBUCK
3½ ft. at shoulder

GREATER KUDU
4½ ft. at shoulder

WOLVES, DOGS, AND FOXES are members of the dog family and are native to all continents. The Dingo, the only wild dog in Australia, was probably brought there by settlers.

Wolves, which once hunted over much of the Northern Hemisphere, are now found only in a few wilderness areas. Coyotes of North America and Jackals of Asia and Africa are smaller than wolves. All of these animals generally hunt in packs. There are many kinds of foxes in the world. Most are smaller than dogs and wolves and do not hunt in packs. The Kit Fox is a small desert species of Western North America.

DINGO
2 ft. tall

KIT FOX
1 ft. tall

COYOTE
1½ ft. tall

WALRUS
10-12 ft., 3,000 lbs.

SEA LION
8-10 ft.,
1,500 lbs.

FUR SEAL
6 ft.

WALRUSES, SEALS, AND SEA LIONS are sea-going flesh eaters, related to bears, racoons, and dogs but not to whales. Some kinds come to land only to breed. They swim with their front and rear flippers. The fur of some is valuable. Fish is their main food. The Walrus digs clams and crabs out of bottom mud in water as deep as 300 feet. The Walrus is found only in the Arctic, but other kinds of seals are found in all oceans. Most Sea Lions as well as the Elephant Seal live in the Pacific. Circus seals are California Sea Lions.

BEARS   All bears except the Spectacled Bear of South America are found in the Northern Hemisphere. The Polar Bear (p. 4) lives in the arctic regions. Black Bears weigh about 300 pounds, and the Brown Bears, which include the Grizzly, may weigh 1,500 pounds or more. Species of Brown Bears are found in Europe, Asia, and North America. Bears do not hibernate in winter, though they may sleep for long periods.

**GRIZZLY
BEAR**
6-7 ft. tall

WOLVERINE
3 ft.

RIVER OTTER
2-2½ ft.

SEA OTTER
5 ft.

HOGNOSED SKUNK
19 in.

BADGER
28 in.

WEASELS, OTTERS, SKUNKS, AND BADGERS Members of the weasel family are found over all the world except in Australia and New Guinea. All have scent glands. Otters are sleek, swift swimmers. The Wolverine of the northern wilderness country can kill animals as large as Caribou. Skunks live in North and Central America, but the very similar Zorilla lives in Africa. Badgers are found in Europe, Asia, and North America.

MONGOOSES, CIVETS, AND GENETS from Europe, Africa, and Asia look like a cross between a cat and a weasel. Most of the many kinds live in the forest where they climb after birds, tree rats and other small animals. A few kinds eat fruit. Many have scent glands. The Mongoose is best known for its fearless attacks on snakes. Civets and Genets both hunt at night and are not well known even in their native lands. Civet musk is used in making perfume in the Near East.

AFRICAN CIVET CAT
32 in.

CATS are quite similar in looks and actions. They are found throughout the world. Large ones roar, small ones purr. All except the fleet Cheetah can pull back their claws into sheaths when they are not in use. Among the largest are Lions (Africa and, rarely, in Asia), Tigers (India to Siberia) and Leopards (Africa and Asia). Cheetahs, of Asia and Africa, can run 70 miles an hour for short distances. They are the fastest of all land animals. The African Lion is the strongest of the big cats. It sometimes climbs trees and lolls lazily on the limbs.

AFRICAN LION 8 ft.

male

female

MOUNTAIN LION
6 ft.

The Mountain Lion, also called Puma, Panther, or Cougar, is the largest cat native to North America. It can kill deer and elk. The heavier Jaguar (300 pounds), is a woodland cat found in South and Central America. The slim Jaguarundi, of the same region, is only about two feet long and has a long tail. The Canada Lynx and the smaller Bobcat were once common in the North. Like the Ocelot of Central and South America, they climb trees to catch birds. They also eat small mammals.

SLOW LORIS
18 in.

LORISES, TARSIERS, AND LEMURS are small monkey-like mammals of Asia and the East Indies. Lemurs are found in Madagascar and Africa. Most feed at night.

NEW WORLD MONKEYS, most with long, grasping tails and flat noses, inhabit the warm forests of the American tropics. Marmosets, the smallest, are about the size of a squirrel. Uakaris and Howlers live in the Amazon Basin. Spider and Capuchin monkeys are often kept as pets.

WHITE-EARED MARMOSET
body 10 in., tail 12 in.

HOWLER MONKEY
body 24 in., tail 30 in.

RED UAKARI
body 25 in.,
tail 10 in.

**WHITE-THROATED
CAPUCHIN**
body 15 in., tail 20 in.

OLD WORLD MONKEYS Only one monkey, the Barbary Ape, lives in Europe, on the Rock of Gibraltar. It also lives in Africa. The Rhesus Monkey of India is regarded as sacred by the Hindus.

GORILLA
6 ft. tall

THE GREAT APES have long arms, short legs and no tail. Gibbons and Orangutans are tree-dwellers of South-east Asia; they seldom descend to the ground. Gorillas and Chimpanzees live in the forests of Africa, spending part of the time in trees and part on the ground. They walk on all fours, using the knuckles of their hands for support.

RHESUS MONKEY
30 in.,
tail 25 in.

CHIMPANZEE
5 ft. tall

BARBARY APE
30 in., no tail

GIANT ANT-EATER
8 ft.

ANT-EATERS use their powerful front claws to dig into the nests of termites and ants, catching their prey on their long, sticky tongue. They do not have teeth. Ant-eaters live in Central and South America; scaly ant-eaters live in Asia and Africa.

ARMADILLOS, found only in the New World, are covered with bony plates. Some kinds can roll into a ball for protection.

NINE-BANDED ARMADILLO
2½ ft.

THREE-TOED SLOTH 20 in.
incubating eggs

SLOTHS are slow-moving leaf eaters from the forests of Central and South America. Their normal way of travel is swinging along upside down on the underside of branches. Their long hair often appears greenish because of tiny plants (algae) growing on it.

PLATYPUS, or Duckbill, lives in Australia. It is considered a primitive mammal because the females lay eggs and incubate them.

PLATYPUS 18 in.

suckling young

MANATEE
12 ft., weight 1,300 lbs
nursing young

MANATEES AND DU-GONGS are water-dwelling plant eaters that swim with their broad tail and flipper-like front legs. They lack hind legs. Manatees, or Sea Cows, live on the tropical Atlantic coast, Dugongs in the Indian Ocean and western Pacific.

QUIZ-ME

Here are some questions you can answer if you have studied this book. The pages where the answers will be found are listed at the end.

1 What kind of animal was probably the ancestor of all mammals?
2 What is the difference between hares and rabbits?
3 What South American mammal is a close relative of the camel?
4 What is the largest toothed whale?
5 What full-grown mammal weighs less than an ounce?
6 How many kinds of wild dogs are found in Australia?
7 Which of the Great Apes live in Africa? In Asia?
8 What do bats eat besides insects?
9 What is the only mammal that can truly fly?
10 Why is the Platypus considered a primitive mammal?
11 Which of these are members of the weasel family— skunk, badger, otter?
12 What mammal travels upside down?
13 Name three kinds of rodents.
14 What is the smallest monkey?

15 What kind of teeth do ant-eaters have?
16 Where do the New World Monkeys live?
17 What do the teeth of elephants and rats have in common?
18 What is the only pouched mammal of N.A.?
19 Do bears hibernate?
20 What does a Walrus eat?
21 What is the proper name for the North American Buffalo?
22 Which cat is the fastest of all animals?
23 How far can a kangaroo jump?

ANSWERS: 1(p. 7), **2**(p. 19), **3**(p. 25), **4**(p. 15), **5**(p. 3), **6**(p. 32), **7**(p. 42), **8**(p. 12), **9**(pp. 4, 12), **10**(p. 45), **11**(p. 36), **12**(p. 45), **13**(pp. 16, 17, 18), **14**(p. 40), **15** (p. 44), **16**(p. 40), **17**(pp. 16, 20), **18**(p. 8), **19**(p. 35), **20**(p. 34), **21**(p. 26), **22**(p. 38), **23**(p. 9).

ILLUSTRATIONS BY: *Dorothea and Sy Barlowe, Arch and Miriam Hurford, James Gordon Irving, Bob Kuhn, Harry McNaught, Don Ray, William de J. Rutherfoord, Arthur Singer.*
COVER BY: *Hal McIntosh*